English
Français
Deutsche
Italiano
Español
Português

www.forgottenbooks.com

Mythology Photography **Fiction**
Fishing Christianity **Art** Cooking
Essays Buddhism Freemasonry
Medicine **Biology** Music **Ancient
Egypt** Evolution Carpentry Physics
Dance Geology **Mathematics** Fitness
Shakespeare **Folklore** Yoga Marketing
Confidence Immortality Biographies
Poetry **Psychology** Witchcraft
Electronics Chemistry History **Law**
Accounting **Philosophy** Anthropology
Alchemy Drama Quantum Mechanics
Atheism Sexual Health **Ancient History**
Entrepreneurship Languages Sport
Paleontology Needlework Islam
Metaphysics Investment Archaeology
Parenting Statistics Criminology
Motivational

ISBN 978-1-333-37509-6
PIBN 10496781

Reprinted from Soil Science
Vol. XII, No. 2, August, 1921

$S573$
$S75$

SOIL ACIDITY AND BACTERIAL ACTIVITY[1]

R. E. STEPHENSON

Agricultural College, University of Kentucky

Received for publication February 17, 1921

INTRODUCTION

'Just how and why soils become acid is a problem that has not yet been definitely solved. Neither is the effect of reaction upon the activity of soil organisms clearly understood. But is has been fairly well established that the process of nitrification once thought to be absent in acid soils, does proceed to an appreciable extent. In fact nitrification is perhaps sufficient for normal crop production, in most cases, provided the supply of organic matter is adequate. The process of ammonification which of course must precede nitrification is carried on by so many classes of organisms that it is not usually a limiting factor in crop production in either acid or sweet soils, under aerobic or anaerobic conditions.

In practically all soils there must be two analytical processes, the decomposition of organic matter, and the disintegration of minerals. The importance of microorganisms in bringing about these processes is too obvious to need comment. While these processes are occurring, plant growth also takes place. The general tendency of plant growth has been found to be to keep the nutrient solution nearly neutral. Crop production therefore doubtless has a tendency to prevent soils from becoming acid in reaction, while the leaching of bases has the opposite effect.

The cultivation of soils for crop production of course encourages leaching, stimulates bacterial activity, and on the whole in this indirect way must tend to produce acid soils.

In mineral disintegration, with the accompanying interchange of ions, both acids and bases must be set free. Similar effects are produced when organic matter is broken down. But changes in the organic portion of the soils must occur under favorable conditions much more rapidly than changes in the mineral portion. The acids and carbon dioxide produced in organic decay hasten mineral disintegration, and therefore increase the availability of mineral

---

[1] Part of the results of this study on "Soil Acidity and Bacterial Activity" has already been published. Two papers, "The Effect of Organic Matter on Soil Reactions. I," and "The Activity of Soil Acids" were published in Soil Science (6, 7), another paper "Nitrification in Acid Soils" is in press at the Iowa Agricultural Experiment Station. This paper is the first part of a thesis presented to the graduate faculty of the Iowa State College of Agriculture in partial fulfilment of the requirements for the degree of Doctor of Philosophy.

plant-food. But though minerals are put into solution by these processes, there is also a compensating effect, in that organic decomposition products are capable of forming insoluble compounds with the minerals disintegrated and thus may prevent or at least retard the loss of the minerals by leaching.

One fact to be kept in mind in connection with both organic acids and with bases, is that so far as available data indicate, these compounds do not remain long as such. Oxidation changes convert the nitrogen bases to nitric acid and the organic acids to carbon dioxide. Only the nitric acid produced, therefore, is capable of causing a permanent direct effect upon soil reaction. Mineral bases and acids, on the other hand, are permanently removed from the soil primarily by leaching. The portion used by the plant may be expected to be returned to the soil, at least in part.

It may be observed, too, that practical experience demonstrates that soils containing sufficient organic matter remain more productive for a longer time than those soils which are low in organic matter. Loss of organic matter is likely to result in a sour, soggy, infertile soil, which does not respond to tillage or commercial fertilizer. Muck and peat soils are notable exceptions but largely because mineral elements, such as potassium and other bases, were never present. And again, such soils occur only under those conditions which favor a large production of organic acids, and prevent complete oxidation. These soils, therefore, are often highly acid, and this condition is undoubtedly due mainly to organic acids. But by way of contrast it must be observed that sandy soils and heavy clays, which do not contain sufficient organic matter to produce an appreciable acidity, are often highly acid and nonproductive.

In this work one heavy silt loam soil, one soil somewhat sandy, both low in organic matter, and a loam soil rather high in organic matter were used, for the purpose of studying the changes which occur, the rate of change, and to some extent the final products of the reactions.

HISTORICAL

Previous investigations of the effect of organic matter upon the reaction of soils is very limited in amount and application. White (8), Skinner and Beatty (3), Miller (2) and Stephenson (6) found no positive evidence that the decay of organic matter in ordinary soils under conditions which would be favorable to crop production, produced any appreciable increase in the lime requirement of the soil. No argument is necessary, of course, to establish the fact that the large production of nitric acid would increase the acid reaction of the soil or use up bases rapidly if they were present.

THE PLAN OF THE EXPERIMENT

In a previous publication (6) the effect of the decomposition of albumin, casein, starch, blood, dextrose, alfalfa, and ammonium sulfate on the reaction of two soils was studied. Further work along this same line is reported here, with organic materials of more general use such as farm manure, cottonseed meal, horse manure, timothy hay, clover hay, green timothy and green clover. Opportunity is thus afforded for comparing the green and the more matured dried materials.

Two of the same soils used in the earlier work were employed, one rather sandy and light in color, the other dark and fairly rich in organic matter, and of the loam type. Applications of the various materials were made at the rate of 10 tons per acre of air-dried material, on the basis of 2,000,000 pounds of soil per acre. The coarse materials were ground and thoroughly mixed with the soils, in 1-gallon earthenware jars. Samplings were made at intervals of 2, 5, 10, 15, and 22 weeks, respectively. Two series were run, one limed and the other unlimed. Determinations were made at each sampling for the ammonia, nitrates, acidity, and residual carbonates, since these are directly connected with the effect of materials on the soil reaction. A test was made at the second sampling, for the soluble non-protein nitrogen present in one of the soil types. This test should throw some light on the question of the possibility of any accumulation of soluble products of protein decomposition, other than nitrates and ammonia, and should also show whether there is any correlation between these products and the quantity of nitrates or ammonia present in soils.

AMMONIFICATION

The quantity of ammonia was determined by the aeration method, potassium carbonate being used to liberate the ammonia. Incidentally it may be said that experience at the Iowa Agricultural Experiment Station with this method would lead to the conclusion that those workers who have found the method unsatisfactory, must have experienced a faulty manipulation. The secret of successful operation of the method, is that the aeration must stir the soil completely to the bottom of the containing flask. The results of the ammonia determinations are given in table 1.

It may be observed that there is very little accumulation of ammonia with any of the treatments except the cottonseed meal. It has shown the greatest accumulation of ammonia at the first sampling and a greater accumulation when the soil was untreated, than when it was limed, both of which results agree with work done previously (6) with highly nitrogenous materials. There is too small an accumulation of ammonia on the untreated soils to show marked differences between the limed and the unlimed soils. The same may be said of most of the other treatments, though there is a greater amount of ammonia in the unlimed soils where green manures were added. The greatest amount of ammonia is found in nearly all cases at the first sampling before nitrification is well started. There is quite a marked difference in the two soils, noticeable where the cottonseed meal is used, in that the amount of ammonia throughout the test remains high on the unlimed sandy soil, while on the humus soil nitrification seems to have just about kept pace with ammonification even in the absence of lime. This result lends support to the belief that soils containing sufficient organic matter are more active bacteriologically, and likewise usually more productive, than soils containing less organic matter even when the total time requirement is much greater for the organic soils.

The amount of ammonia produced may depend upon several factors. But when conditions are favorable for nitrification the ammonia is changed to nitrates almost as rapidly as produced.

TABLE 1

Amount of ammonia at the end of each period

TREATMENT	FIRST SAMPLE, 2 WEEKS		SECOND SAMPLE, 5 WEEKS		THIRD SAMPLE, 10 WEEKS		FOURTH SAMPLE, 15 WEEKS		FIFTH SAMPLE, 22 WEEKS		AVERAGES	
	No lime	Lime	No lime	Lime	No lime	Lime	No lime	Lime	No lime	Lime	No lime	Lime
	p.p.m.	p.p.m.	p.p.m.	p.p.m.	p.p.m.	p.p.m.	p.p.m.	p.p.m.	p.p.m.	p.p.m.	p.p.m.	p.p.m.
*H*umus soil:												
Soil alone.....	11.8	14.0	11.2	8.4	16.8	11.2	11.2	14.0	11.2	8.4	13.4	11.1
Cottonseed meal.......	302.4	285.6	268.8	61.6	98.0	22.4	86.8	19.6	32.0	14.0	157.6	80.6
Manure......	8.4	5.6	11.2	8.4	11.2	11.2	14.0	11.2	14.0	11.2	11.7	9.6
Timothy hay	5.6	8.4	11.2	11.2	16.8	11.2	11.2	5.6	11.2	11.2	11.2	9.5
Clover hay...	19.6	11.2	8.4	5.6	11.2	11.2	5.6	11.2	11.2	11.2	11.2	10.1
Green timothy.	44.8	11.2	14.0	5.6	16.8	11.2	11.2	11.2	11.2	11.2	19.6	10.1
Green clover..	33.6	14.0	16.8	5.6	16.8	11.2	11.2	8.4	11.2	8.4	17.9	9.5
Average....	61.6	50.0	48.8	15.2	22.8	12.8	21.6	11.6	14.6	10.8	48.5	28.1
Sandy soil:												
Soil alone......	56.0	30.8	14.0	5.6	16.8	11.2	19.6	11.2	14.0	14.0	24.1	14.6
Cottonseed meal.......	294.8	305.2	280.0	100.8	132.5	16.8	151.2	22.4	14.0	19.6	194.5	92.9
Manure......	16.8	19.6	8.4	11.2	8.4	11.2	·8.4	8.4	11.2	11.2	10.6	12.3
Timothy hay..	11.2	8.4	11.2	8.4	16.8	11.2	14.0	11.2	11.2	89.6	12.9	25.8
Clover hay...	39.2	39.2	19.6	11.2	14.0	14.0	11.2	8.4	16.8	14.0	20.1	17.4
Green timothy.	58.8	47.6	33.6	16.8	14.0	8.4	11.2	8.4	5.6	5.6	24.6	17.8
Green clover..	103.6	75.6	39.8	14.0	11.2	11.2	11.2	11.2	5.6	5.6	32.5	23.5
Average....	97.2	75.2	56.8	24.0	30.5	12.0	32.4	11.6	17.2	22.8	45.6	29.2

Lime favors nitrification and at least in that indirect way indicates a retarded ammonification. Lime also increases the number of organisms, and should therefore tend to reduce the total of ammonia and nitrates in the presence of a limited supply of organic matter, because of greater nutritional demands by the increased number of organisms. When a large amount of nitrogenous organic matter is added perhaps this would not result. And since the ammonification process is the actual limiting factor under conditions which permit of nitrification, the increased basicity due to the use of lime evidently does have a retarding effect.

When averages are taken of all determinations and all treatments, there is no case on the humus soil (so-called because of its higher content of organic matter) where lime has not diminished the amount of ammonia produced. On the sandy soil there are two cases, with manure and with timothy hay, where the reverse is true, but the result would appear to be more nearly accidental than fundamental.

TABLE 2

Nitrates at each successive sampling

TREATMENT	FIRST SAMPLE, 2 WEEKS		SECOND SAMPLE, 5 WEEKS		THIRD SAMPLE, 10 WEEKS		FOURTH SAMPLE, 15 WEEKS		FIFTH SAMPLE, 22 WEEKS		AVERAGE	
	No lime	Lime	No lime	Lime	No lime	Lime	No lime	Lime	No lime	Lime	No lime	Lime
	p.p.m.	*p.p.m.*	*p.p.m.*	*p.p.m.*	*p.p.m.*	*p.p.m.*	*p.p.m.*	*p.p.m.*	*p.p.m.*	*p.p.m.*	*p.p.m.*	*p.p.m.*
Humus soil												
Soil alone.....	28.6	19.1	63.5	68.8	38.9	95.9	52.3	102.0	50.0	121.1	64.7	83.4
Cottonseed meal.......	33.0	45.7	98.3	243.2	214.8	309.0	302.4	289.9	324.0	316.0	194.5	240.8
Manure......	14.2	7.3	21.4	23.8	37.8	57.8	36.7	61.8	74.5	104.1	36.9	50.9
Timothy hay.'	Tr.*	Tr.	Tr.	Tr.	Tr.	20.5	Tr.	35.5	22.8	67.4	4.5	24.7
Clover hay...	40.6	58.9	67.8	92.5	80.3	129.5	86.3	133.5	116.7	170.8	78.3	117.0
Green timothy.	45.6	51.5	100.4	83.9	180.5	125.0	141.1	93.8	181.4	121.4	129.8	94.9
Green clover..	69.4	78.1	109.7	122.0	234.1	319.1	181.5	168.1	284.6	201.0	175.8	177.6
Average....	33.1	37.2	65.9	90.6	112.3	150.9	114.3	127.8	150.6	165.9	95.2	112.8
Sandy soil:												
Soil alone.....	17.7	16.6	58.6	72.4	85.0	58.8	97.6	73.1	81.6	103.8	68.1	65.9
Cottonseed meal.......	9.4	7.3	112.2	138.3	167.9	229.4	267.6	400.2	312.4	457.4	173.9	246.5
Manure......	11.2	19.1	38.2	52.1	53.1	62.1	61.4	68.8	61.8	89.8	45.0	54.4
Timothy hay..	Tr.	Tr.	Tr.	Tr.	Tr.	14.8	Tr.	41.5	21.3	50.3	4.2	21.3
Clover hay...	11.5	15.1	63.6	97.1	83.5	69.5	90.7	122.0	123.1	152.4	74.5	91.2
Green timothy.	12.1	23.3	66.1	86.6	100.7	82.4	92.0	88.0	105.3	144.3	75.2	84.9
Green clover..	16.4	13.7	86.0	109.3	153.3	117.9	147.3	135.5	207.4	183.7	122.1	112.0
Average....	11.2	13.6	60.7	79.4	90.0	90.7	108.0	129.9	130.3	168.8	80.4	96.6

* Tr. = trace.

NITRIFICATION

For the determination of nitrates the phenoldisulfonic acid method as modified by Davis (1) was used. Calcium carbonate was employed to flocculate the soil and secure a clear filtrate. The results are given in table 2.

It is observed that the amount of nitrates increased in the untreated soils up to the last sampling.

The cottonseed meal, in accordance with its higher nitrogen content, gave a greater accumulation of nitrates on both soils than any other treatment. Here again, the sandy soil, though starting more slowly, finally ran higher than the better soil. On both soils, the greatest amount of nitrate was found at the last sampling, the first two samples on the sandy soil showing less than the untreated soil. In most cases lime increased the nitrification of cottonseed meal.

The addition of stable manure caused a decrease in the amount of nitrates present in most cases, probably because of an increased number of organisms greater than the accompanying addition of easily nitrifiable material.

Timothy hay had the same effect as stable manure but to a much more marked degree. Little nitrifiable material was added in the timothy, but considerable energy material was provided, and the organisms used most of the nitrates for nutritional purposes. The nitrates began to show at about the same time on both soils but never ran nearly so high as on the untreated soils. Lime again stimulated nitrification. The green timothy in contrast to the dry, stimulated nitrification at once on both soils, and the greatest accumulation of nitrates was found at the last sampling and in the presence of lime.

Dry clover also caused a gradual stimulation of nitrification, the greatest effect being produced at the last sampling. The stimulation was usually greater also in the presence of lime. The green clover had a somewhat greater effect than did the dry, and maximum nitrification was induced sooner.

When averages of all samplings and all treatments are taken, the humus soil shows greater nitrification in the presence of lime in every case except one, and this is where green timothy was applied. There is very little difference with the green clover. When the sandy soil is considered the soil alone produces slightly less nitrates on the limed series. Every treatment except one, and in this case it is green clover, has shown greater nitrification in the presence of lime. Apparently the lime does not affect the nitrification of the green material as much as some of the dried materials. As is quite logical, the greatest amount of nitrates is found at the last sampling, while the greatest amount of ammonia is usually found at the first sampling.

A summary of the nitrate and ammonia determinations is given in table 3.

The table shows the largest combined production of nitrates and ammonia where cottonseed meal was applied, followed in order by green clover, green timothy, horse manure, and dry timothy, the latter two producing considerably less than the soil alone. The general effect of the lime was to decrease the total of nitrates and ammonia found, especially where there was any large production.

ACIDITY RESULTS

The lime requirements on the soils differently treated are given in table 4. The determinations were made according to the modified Tacke method previously described (5). The acid soil was brought into contact with pure calcium carbonate, and the aeration and shaking continued for 10 hours before titrations were made. The double-titration was performed, with methylorange and phenolphthalein as indicators.

TABLE 3

Nitrogen summary, summary of ammonia and nitrates.

TREATMENT	FIRST SAMPLE 2 WEEKS		SECOND SAMPLE 5 WEEKS		THIRD SAMPLE 10 WEEKS		FOURTH SAMPLE 15 WEEKS		FIFTH SAMPLE 22 WEEKS		AVERAGE			
											No lime		Lime	
	No lime p.p.m.	Lime p.p.m.	No lime p.p.m.	Lime p.p.m.	No lime p.p.m.	Lime p.p.m.	No lime p.p.m.	Lime p.p.m.	No lime p.p.m.	Lime p.p.m.	p.p.m.	Minus soil p.p.m.	p.p.m.	Minus soil p.p.m.
Humus soil:														
Soil alone	45.5	33.1	74.7	77.3	55.7	107.2	63.6	131.1	61.2	134.5	60.1		96.6	
Cottonseed meal	335.5	331.3	367.1	304.8	312.9	331.4	338.2	309.5	346.1	330.0	339.9	279.8	321.4	224.8
Manure	22.6	13.0	32.6	33.3	49.3	69.0	50.8	73.0	88.5	115.3	48.7	−11.4*	60.7	−35.9*
Timothy hay	5.6	8.4	11.2	11.2	16.8	31.6	11.2	41.1	34.0	78.6	15.7	−44.4*	34.2	−62.4*
Clover hay	79.2	70.1	76.5	98.1	91.5	140.6	91.9	144.7	127.7	170.8	93.4	33.3	124.8	28.2
Green timothy	90.4	62.7	114.4	89.5	197.3	136.1	152.2	105.0	192.6	132.6	149.4	89.3	105.2	8.6
Green clover	103.1	92.1	125.5	127.6	250.9	330.3	192.7	176.5	295.8	209.4	193.8	133.7	187.2	90.6
Average											128.7	134.0	132.9	88.1
Sandy soil:														
Soil alone	73.8	37.4	72.6	78.0	101.8	70.0	117.2	84.8	95.6	117.8	92.2		77.6	
Cottonseed meal	404.2	312.5	392.2	239.1	300.4	246.2	418.8	422.6	326.4	477.0	368.4	276.2	339.5	261.9
Manure	28.0	38.7	46.6	63.4	61.5	73.7	69.5	57.2	72.6	101.0	55.6	−36.6*	66.8	−10.8*
Timothy hay	11.2	8.4	11.2	8.4	16.8	26.1	14.0	52.7	32.5	139.9	17.1	−75.1*	47.1	−3.0*
Green hay	50.8	54.3	143.2	107.3	97.6	83.5	101.9	130.4	139.8	171.4	106.7	14.5	109.4	31.8
Green timothy	70.9	70.9	99.7	103.4	114.7	90.8	103.2	101.4	110.9	149.9	99.9	7.7	103.3	25.7
Green clover	114.0	89.3	116.8	123.3	164.5	129.1	158.5	146.7	212.6	189.3	153.3	61.1	135.5	57.9
Average											127.6	89.9	125.6	94.3

* These are omitted in taking final averages.

There is little to be said in regard to the effect of the various treatments upon the lime requirement of the soils. The general tendency has been to

TABLE 4

Lime requirement of the variously treated soils in tons per 2,000,000 pounds soil

TREATMENT	FIRST SAMPLE, 2 WEEKS	SECOND SAMPLE, 5 WEEKS	THIRD SAMPLE, 10 WEEKS	FOURTH SAMPLE, 15 WEEKS	FIFTH SAMPLE, 22 WEEKS	MORE OR LESS THAN THE SOIL ALONE				
						First sample	Second sample	Third sample	Fourth sample	Fifth sample
	tons	tons	tons	tons	tons	tons	tons	tons	tons	tons
*H*umus soil:										
Soil alone..............	3.90	4.20	3.85	3.80	3.80					
Cottonseed meal.........	3.65	3.65	4.45	4.25	4.55	−0.25	−0.55	+0.60	+0.45	+0.75
Manure.................	3.80	4.25	3.60	3.40	3.80	−0.10	+0.05	−0.25	−0.40	+0.00
Mature timothy..........	4.05	4.15	3.55	3.35	3.55	+0.15	+0.05	−0.30	−0.45	−0.25
Mature clover...........	3.35	4.15	3.65	3.25	3.95	−0.05	−0.05	−0.20	−0.55	+0.15
Green timothy...........	4.10	4.45	3.70	3.65	3.95	+0.20	+0.25	−0.15	−0.15	+0.15
Green clover............	3.6	4.00	3.25	3.20	3.85	−0.20	−0.20	−0.60	−0.60	+0.05
Sandy soil:										
Soil alone..............	3.20	2.60	2.35	2.40	2.35					
Cottonseed meal.........	1.70	2.15	2.15	2.50	2.45	−0.50	−0.45	−0.20	+0.10	+0.10
Manure.................	2.20	2.35	2.10	2.65	1.75	0.00	−0.25	−0.25	+0.25	−0.60
Mature timothy.........	2.20	2.30	1.80	2.05	1.75	0.00	−0.30	−0.55	−0.35	−0.60
Mature clover...........	2.15	2.30	1.90	1.80	1.75	−0.05	−0.30	−0.45	−0.60	−0.60
Green timothy...........	2.55	2.65	2.25	2.30	2.00	+0.35	+0.05	−0.10	−0.10	−0.35
Green clover............	1.70	2.65	1.90	1.90	1.85	−0.50	+0.05	−0.45	−0.50	−0.50

TABLE 5

Difference of ammonia and nitrates on unlimed soils compared with effect of treatment on lime requirement

*H*umus soil:					
Ammonia (p.p.m.)...............	302.4	268.8	98.0	86.8	32.0
Nitrates (p.p.m.)................	33.0	98.3	214.8	302.4	324.0
Difference (p.p.m.)...............	+269.4	+170.5	−116.8	−215.6	−292.0
Difference in lime requirement (tons)........................	−0.25	−0.55	+0.60	+0.45	+0.75
Sandy soil:					
Ammonia (p.p.m.)...............	394.8	280.0	132.5	151.2	14.0
Nitrates (p.p.m.)................	9.4	112.2	167.9	267.6	312.4
Difference (p.p.m.)...............	+385.4	+168.8	−35.4	−116.4	−298.4
Difference in lime requirement (tons)........................	−0.50	−45.0	−0.20	+0.10	+0.10

reduce rather than to increase it. A large production of ammonia reduces the lime requirement, and, quite logically, when nitrification occurred the opposite effect resulted.

Table 5 brings out this point when the cottonseed meal treatment is studied, in comparing the effect of ammonification and nitrification upon the decrease or increase of the lime requirement of the treated soil over the untreated.

This table shows that though there is not a close correlation between the difference of ammonia and nitric acid produced on the soils treated with cottonseed meal, and the effect upon the lime requirement, the tendency is for the soil to show a greater or smaller lime requirement according as there is more or less of the nitrogen present in the basic or acid form. None of the other treatments contain sufficient nitrogen to make the comparison significant.

TABLE 6

Residual carbonates on treated soils; expressed in tons per acre

TREATMENT	FIRST SAMPLE, 2 WEEKS	SECOND SAMPLE, 5 WEEKS	THIRD SAMPLE, 10 WEEKS	FOURTH SAMPLE, 15 WEEKS	FIFTH SAMPLE, 22 WEEKS	MORE OR LESS THAN SOIL UNTREATED				
						First sample	Second sample	Third sample	Fourth sample	Fifth sample
	tons	tons	tons	tons	tons	tons	tons	tons	tons	tons
Humus soil:										
Soil alone..............	3.40	2.55	2.00	1.95	1.35					
Cottonseed meal.........	4.95	2.55	1.20	1.25	0.55	+1.55	+0.00	−0.80	−0.70	−0.80
Manure.................	4.10	2.85	2.45	1.90	2.05	+0.70	+0.30	+0.45	−0.05	+0.70
Dry timothy.............	4.35	2.90	2.35	2.10	1.90	+0.95	+0.35	+0.35	+0.15	+0.55
Dry clover..............	4.15	3.05	2.30	2.10	2.15	+0.75	+0.50	+0.30	+0.15	+0.80
Green timothy...........	4.05	3.20	2.50	2.30	2.15	+0.65	+0.65	+0.50	+0.35	+0.80
Green clover............	4.20	3.00	2.95	2.50	2.45	+0.80	+0.45	+0.95	+0.55	+1.10
Sandy soil:										
Soil alone..............	2.80	2.55	2.45	2.55	2.40					
Cottonseed meal.........	3.90	2.35	1.70	1.20	0.85	+1.10	−0.20	−0.75	−1.35	−1.55
Manure.................	2.95	2.70	2.65	2.60	2.40	+0.15	+0.15	+0.20	+0.05	+0.00
Dry timothy.............	3.25	2.75	2.45	2.60	2.35	+0.45	+0.20	+0.00	+0.05	−0.05
Dry clover..............	3.30	3.00	2.75	2.90	2.50	+0.50	+0.45	+0.30	+0.35	+0.10
Green timothy...........	2.80	2.55	2.50	2.40	2.30	+0.00	+0.00	+0.05	−0.15	−0.10
Green clover............	4.25	3.30	3.00	3.00	2.85	+0.45	+0.75	+0.55	+0.45	+0.45

RESIDUAL CARBONATES

The residual carbonates were determined by decomposing the remaining limestone with dilute acid, and titrating the carbon dioxide liberated, in the same way as the titration was made in the lime-requirement determinations. The results are given in table 6.

Lime was applied at the rate of 7 tons on the more acid soil and 6 tons on the other soil, in the form of the precipitated carbonate. As was intended a sufficient excess was added so that nitrification did not exhaust it.

The data show that in most cases the organic materials' have tended to protect the lime applied to the soil. The notable exception is the cottonseed meal, which on account of the large production of nitric acid, has used up the carbonates nearly completely. All of the treatments helped to save limestone until nitrification occurred, as noted by the fact that with but three exceptions minus quantities do not appear until the last two samplings.

SOLUBLE NON-PROTEIN NITROGEN

The method employed in this study was in general that used by Potter and Snyder (4). The soil was extracted with 1 per cent hydrochloric acid, in both the limed and the unlimed series. The nitrate nitrogen and the ammonia nitrogen were distilled off by the Devarda reduction method. The residue from this reduction was then treated with sulfuric acid and the total nitrogen determined in the usual way. This latter gave the unknown soluble non-protein nitrogen of the acid extract.

The acid-extracted soil was next extracted by shaking 2 hours with 1.75 per cent sodium hydroxide, and the extract clarified by centrifuging for 5

TABLE 7

Soluble non-protein nitrogen in humus soil after 5 weeks

TREATMENT	UNKNOWN NON-PROTEIN NITROGEN				TOTAL UNKNOWN NON-PROTEIN NITROGEN	
	In HCl extract		In alkaline extract			
	No lime	Lime	No lime	Lime	No lime	Lime
	p.p.m.	*p.p.m.*	*p.p.m.*	*p.p.m.*	*p.p.m.*	*p.p.m.*
Soil alone...............	23.33	26.00	245.5	246.5	268.83	272.50
Cottonseed meal........	195.99	44.66	310.5	287.5	506.49	355.16
Manure.................	32.66	11.33	246.5	218.0	278.16	229.33
Timothy...............	28.66	29.99	232.0	253.0	260.66	282.99
Clover.................	23.33	26.00	260.5	244.0	283.83	270.00
Green timothy...........	35.33	30.00	266.0	253.0	301.33	283.00
Green clover............	28.66	19.60	277.3	253.3	305.96	272.90

minutes at 30,000 revolutions per minute. The extract was then neutralized with sulfuric acid, and acidified with tri-chlor-acetic acid sufficiently to give $2\frac{1}{2}$ per cent of the latter. The precipitate was then filtered off and another aliquot of the filtrate taken for determination of the nitrogen by the micro-method.

Soluble non-protein materials should probably be the largest in amount when decomposition is the most active. But the question is, do these compounds, many of which are doubtless of a peptide character, tend to accumulate in soils in appreciable amounts, or do ammonification and nitrification proceed at once when the decomposition has started. In other words, should the soluble nitrogen be found primarily in the form of ammonia and nitrates or also in more complex forms? Previous study has shown that plants are capable of using more complex forms of nitrogen than nitrates and ammonia, and if they occur to any extent in ordinary soils, there may be conditions when such complex compounds function as direct sources of plant-food.

The results show in every case but one (timothy) that the application of lime has diminished the total unknown soluble non-protein nitrogen. The nitrates and ammonia, though soluble non-protein nitrogen, are not included in these data. A reference to table 3 shows that this is the same general tendency as observed in the production of ammonia. There is one noticeable fact, and that is that none of the organic treatments have as marked an effect upon the amount of unknown soluble non-protein nitrogen as they have on the nitrates and ammonia. This indicates, as do also the data of Potter and Snyder (4), that in the decomposition of proteins of the soil the degradation products undergo rather rapidly a complete change to the simpler state of ammonia. and nitrate. Except in case of the more resistant forms, possibly polypeptides of some degree of complexity, the products apparently do not accumulate to a large extent, and the nitrogen of the soil must exist mostly as the more complex and resistant forms or else as the simplest possible products of decomposition. Ordinarily, of course, nitrates and ammonia are removed from the soil almost as rapidly as produced, and therefore they are not found in large amounts at any one time. Hence the soluble non-proteins such as are found in this study are probably present at any definite time in perhaps five or even ten times the amount of ammonia and nitrates present.

Another question to consider is the possible effect of such compounds on the reaction of the soil. Though perhaps capable of reaction as either acids or bases, they are not found in sufficient quantity to exert a marked effect upon soil reaction. Such materials and others, however, doubtless exercise a buffering effect and help to reduce the hydrogen-ion concentration to some extent.

GENERAL DISCUSSION

This experiment was continued for 159 days, or about 22 weeks. It is not presumed that there would be no change after this time, but rather that such changes as occurred previous to this would determine whatever effects were to be produced by the different treatments on the activity of soil organisms or the reaction of the soil.

The materials used contained the following percentages of nitrogen: dry timothy, 0.693; manure, 1.24; green timothy, 1.28; dry clover, 2.30; green clover, 2.82; and cottonseed meal, 6.96 per cent. The poorer soil contained 0.116, and the better soil about twice as much, or 0.238 per cent, of nitrogen. The amounts of nitrogen found as ammonia and nitrates were for the most part in the same order as the percentages of nitrogen contained in the materials used.

No definite conclusions may be drawn from a limited study, but in general it seems that the essential soil organisms are active in soils of at least moderately strong acidity. The data indicate also that the decay of organic materials under aerobic conditions does not produce an appreciable acidity except where nitric acid is formed in nitrification.

SUMMARY

1. The lime requirement of neither soil was increased by the organic treatments except in those cases where there was a large production of nitric acid.

2. Ammonification is apparently greater in the absence of lime, partly perhaps because of the fact that nitrifying organisms have been less active.

3. Lime has generally stimulated nitrification.

4. The sum of ammonia and nitrates is usually greater on the unlimed soil when treated with nitrogenous organic materials. This is doubtless partly due to the increased number of organisms in the presence of lime and the consequent greater consumption of nitrates and ammonia by the organisms.

5. When nitrogenous sources of energy such as horse manure and timothy hay were supplied, nitrifiction and ammonifiction were reduced below that of the untreated soil.

6. The green materials were somewhat more readily attacked than the dried materials. There was greater production of ammonia and nitrates partly however because of the fact that these materials were richer in nitrogen than the mature plants.

7. The soluble unknown non-protein nitrogen determined at the second sampling on the more fertile soil, when the activity of the organisms was nearly at the maximum, showed little effect due to the various organic treatments. The cottonseed meal was the only treatment which gave any large increase over the untreated soil.

8. In all cases but one, the unlimed treatments gave a higher non-protein nitrogen content than the limed.

REFERENCES

(1) DAVIS, C. W. 1917 Studies on the phenol-disulphonic acid method for determining nitrates in soils. *In* Jour. Indus. Engin. Chem., v. 9, no. 3, p. 290.

(2) MILLER, M. F. 1917 Effect of the addition of organic matter to the soil upon the development of soil acidity. *In* Mo. Agr. Exp. Sta. Bul. 147, p. 50–51.

(3) SKINNER, J. J., AND BEATTIE, J. H. 1917 Influence of fertilizers and soil amendments on soil acidity. *In* Jour. Amer. Soc. Agron., v. 9, no. 1, p. 25–35.

(4) SNYDER, R. S., AND POTTER, R. S. 1918 Soluble non-protein nitrogen of the soil. *In* Soil Sci., v. 6, p. 441–448.

(5) STEPHENSON, R. E. 1918 Soil acidity methods. *In* Soil Sci., v. 6, p. 33–52.

(6) STEPHENSON, R. E. 1918 The effect of organic matter on soil reaction. *In* Soil Sci., v. 6, p. 413–439.

(7) STEPHENSON, R. E. 1919 The activity of soil acids. *In* Soil Sci., v. 8, p. 41–59.

(8) WHITE, J. W. 1918 Soil acidity and green manures. *In* Jour. Agr. Res., v. 13, no. 3, p. 171–197.

Reprinted from SOIL SCIENCE
Vol. XII, No. 2, August, 1921

THE EFFECT OF ORGANIC MATTER ON SOIL REACTION. II[1]

R. E. STEP*H*ENSON

Agricultural College, University of Kentucky

Received for publication February 17, 1921

INTRODUCTION

A study on the effect of organic matter on soil reaction was undertaken as a part of an extended investigation of soil acidity. For a description of the background of the experiments here reported, experimental methods, etc., the reader is referred to the preceding study (5), also to a former study of the same problem (4).

In this series of treatments the organic materials were applied at the same rates as before (10 tons) (4) except where dried blood and straw were mixed and then blood was used at the rate of 10 tons, with 5 and 10 tons of straw. Precipitated carbonate of lime was added to the limed treatments at the uniform rate of 5 tons per acre. The materials used were soybean hay, green rape, oat straw, green soybean hay (pods removed), dried blood and a mixture of blood and oat straw, all in both the limed and the unlimed series. The green materials were dried, as were also the other materials, and ground as finely as was convenient before adding to the soil. The soil used in this study was an acid silt loam taken from the West Virginia Agricultural Experiment Station farm, rather heavy and compact, and poor in organic matter.

The total period of incubation was 125 days, samplings being made at intervals of 2, 5, 10 and 18 weeks, respectively. In addition to the determinations made in the study of the previous series, hydrogen-ion determinations were made upon all treatments.

AMMONIFICATION

The aeration method was again used for ammonia. The results are shown in table 1, expressed as parts of nitrogen per million of soil.

Only the blood possessed a high nitrogen content and therefore it is the only material which caused a large production of ammonia.

[1] This paper is the second part of a thesis presented to the graduate faculty of the Iowa State College of Agriculture in partial fulfillment of the requirements for the degree of Doctor of Philosophy. It is also the second paper published on this study, the former (4) having appeared in 1919. A portion of the work here reported was completed at the Iowa Agricultural Experiment Station, and the remainder was conducted in consultation with Prof. R. M. Salter at West Virginia University. Acknowledgments are extended to Dr. P. E. Brown, of Iowa State College, and also to Professor Salter, for helpful suggestions in planning and interpreting the work.

Lime produced no marked effect in the ammonification of any of the materials until the third sampling, when it caused an appreciable reduction, which was still very evident at the last sampling. This may have been due to two causes. The lime may have caused greater numbers of organisms to grow, which in turn caused a greater consumption of ammonia. The principal cause, no doubt, was that lime permitted greater nitrification, and most of the ammonia had been changed over to nitrate. The data show that this had occurred.

The oat straw depressed ammonification just as it did nitrification, in most cases below that of the untreated soil; this would indicate that it was a suitable source of energy for bacterial activity.

Green soybeans likewise depressed ammonification below that of the soybean hay but partly because of the fact that their nitrogen content was lower.

TABLE 1

Amount of ammonia at the end of each period

TREATMENT	FIRST SAMPLE, 2 WEEKS		SECOND SAMPLE, 5 WEEKS		THIRD SAMPLE, 10 WEEKS		FOURTH SAMPLE, 18 WEEKS		AVERAGES	
	No lime	Lime	No lime	Lime	No lime	Lime	No lime	Lime	No lime	Lime
	p.p.m.	*p.p.m.*	*p.p.m.*	*p.p.m*	*p.p.m.*	*p.p.m.*	*p.p.m.*	*p.p.m.*	*p.p.m.*	*p.p.m.*
Silt loam soil:										
Soil alone...............	3.6	50.0	32.0	34.0	60.0	5.6	16.0	12.0	36.0	2.54
Soybean hay............	94.0	106.0	107.0	106.0	92.4	8.7	36.0	8.0	82.4	57.2
Green rape..............	182.0	168.0	178.0	132.0	36.4	8.7	64.0	4.0	115.1	78.2
Green soybeans..........	48.0	63.0	40.0	56.0	16.9	22.2	12.0	8.0	29.2	38.5
Oat straw...............	52.0	20.0	24.0	16.0	13.8	5.6	10.4	8.0	25.1	12.4
Blood..................	342.0	282.0	566.0	425.0	546.0	361.2	328.0	54.0	445.5	287.3
Blood and 5 tons of straw..	242.0	316.0	424.0	400.0	336.2	43.2	440.0	48.0	360.5	201.8
Blood and 10 tons of straw.	226.0	300.0	396.0	306.0	288.4	26.8	366.0	32.0	319.1	166.2
Averages...............	169.4	180.0	247.9	209.7	190.1	68.1	179.5	23.1	196.7	120.2

Green rape, on the other hand, stimulated ammonification next to the dried blood. However, it contained a little less nitrogen than the soybean hay, though more than the green soybeans.

Straw mixed with blood had little consistent effect upon ammonification. However, the ammonia produced by the combined application of blood and straw was seldom greater and often less than that produced from the blood alone.

There were individual cases where the limed treatments produced more ammonia than the unlimed, but when averages of all treatments (omitting the soil alone) and of all samples, were taken, the unlimed treatments have produced a greater quantity of ammonia. The difference is quite marked at later samplings when nitrification is well under way.

The data show that the accumulation of nitrates has increased at each successive sampling with all treatments, as well as with the untreated soil, in both the limed and the unlimed series.   In general, there has been greater nitrification in the presence of lime.   This is more noticeable after the first sampling and with the nitrogen rich materials.   Lime apparently had the opposite effect where oat straw was used.   Straw used with blood retarded nitrification at first but later there was little or no retardation.   The maximum amount of nitrates occurred at the last sampling in most cases.

Apparently the green soybeans began to nitrify more quickly than did the soybean hay.   Green rape likewise at once stimulated nitrification to an appreciable extent.

NITRIFICATION

Nitrates were determined by the colorimetric method as before.   The results are shown in table 2.

TABLE 2

Nitrates at each successive sampling

TREATMENT	FIRST SAMPLE, 2 WEEKS		SECOND SAMPLE, 5 WEEKS		THIRD SAMPLE, 10 WEEKS		FOURTH SAMPLE, 18 WEEKS		AVERAGES	
	No lime	Lime	No lime	Lime	No lime	Lime	No lime	Lime	No lime	Lime
	p.p.m.	*p.p.m.*	*p.p.m.*	*p.p.m.*	*p.p.m.*	*p.p.m.*	*p.p.m.*	*p.p.m.*	*p.p.m.*	*p.p.m.*
Silt loam soil:										
Soil alone................	19.6	23.0	24.8	33.3	38.4	54.6	65.3	113.8	37.0	56.2
Soybean hay.............	25.9	25.7	28.8	37.9	71.8	181.1	122.6	195.3	62.3	110.0
Green rape..............	76.1	96.6	79.0	75.1	120.5	234.1	260.3	188.7	133.9	148.6
Green soybeans..........	42.5	27.2	388.5	39.7	83.5	108.2	175.7	109.0	172.5	71.0
Oat straw................	19.8	Tr.*	19.6	4.6	26.3	16.2	38.6	38.9	26.1	15.4
Dried blood..............	14.2	9.1	24.5	40.7	149.1	485.5	353.3	611.1	135.3	286.6
Blood and 5 tons of straw.	8.6	Tr.	17.3	37.1	160.3	280.9	332.7	640.0	129.8	264.5
Blood and 10 tons of straw.	Tr.	Tr.	59.1	59.8	156.9	492.5	413.0	575.4	157.2	281.9
Average.................	26.7	22.7	88.2	42.1	109.8	271.5	242.3	336.9	116.7	168.3

* Tr. = trace.

Nitrification apparently scarcely occurred in the presence of oat straw until the third and fourth samplings.   In no case was there as much nitrate as on the untreated soil.

Nitrification was slow in starting when blood and straw were mixed but by the end of 10 weeks there was an appreciable accumulation of nitrates on the treated soils over the untreated.   Apparently the addition of straw had no marked effect upon the nitrification of dried blood.

When averages of all treatments and all samplings are taken (omitting the untreated soil) it is observed that nitrification was slow in starting where straw and blood and mixtures of the two were used, but the blood-straw mixtures finally ran high.   Lime in these cases seems to have retarded the beginning of the nitrifying process, but perhaps more organisms were present where lime was added and they were consuming such nitrates as were produced.

The nitrogen summary shown in table 3 indicates that the average total of nitrates and ammonia has been greatest in most cases for the treated soils, when not limed, but that the reverse is true for the untreated soil. Whether the difference may be due to numbers of organisms and the consequent utilization of part of the nitrogen changed on treated limed soils, cannot be stated, though it seems probable. Experience has shown that in nearly every case a carbohydrate material such as straw which is poor in nitrogen, has given a decrease in ammonia and nitrates over the soil alone, either limed or unlimed. Since the ammonia and nitrate forms of nitrogen are by-products of the attempt of the organism to secure sufficient energy, this is to be expected.

TABLE 3

Nitrogen summary, nitrates and ammonia

TREATMENT	FIRST SAMPLE, 2 WEEKS		SECOND SAMPLE, 5 WEEKS		THIRD SAMPLE, 10 WEEKS		FOURTH SAMPLE, 18 WEEKS		AVERAGE			
	No lime	Lime	No lime	Lime	No lime	Lime	No lime	Lime	No lime		Lime	
	p.p.m.	*p.p.m.*	*p.p.m.*	*p.p.m.*	*p.p.m.*	*p.p.m.*	*p.p m.*	*p.p.m.*	*p.p.m.*	Minus soil	*p.p.m.*	Minus soil
Silt loam soil:												
Soil alone...	55.6	73.0	57.3	67.3	98.4	59.9	81.3	125.8	73.1	*p.p.m.*	81.5	*p.p.m.*
Soybean hay......	120.0	131.8	130.8	144.0	164.2	189.8	158.6	203.3	143.4	70.3	167.2	85.7
Green rape..	258.1	264.6	257.0	207.1	156.9	242.8	324.3	192.7	249.1	176.0	226.8	145.3
Green soybeans .	90.5	95.2	78.5	95.7	100.4	130.4	187.7	117.0	114.3	−41.2	109.8	28.3
Oat straw...	71.8	20.0	43.6	20.6	40.1	23.8	49.1	46.9	51.2	−21.9	27.8	−53.7
Dried blood.	356.2	291.1	590.6	492.7	745.1	846.7	681.1	665.1	580.8	507.7	548.9	467.4
Blood and 5 tons of straw....	250.6	316.0	441.8	437.3	496.5	424.1	777.7	687.9	491.7	418.6	475.8	394.3
Blood and 10 tons of straw....	266.0	300.0	455.0	365.8	445.3	519.2	779.0	607.4	477.2	404.1	488.1	366.6
Average....	196.1	202.7	336.1	251.8	299.9	339.6	421.8	360.0	301.1		286.3	

LIME REQUIREMENT

The data show that in nearly every case the lime requirement was less when organic matter was added to the soil (table 4). The greatest effect was usually at the first sampling. This was especially marked with the dried blood which produced large amounts of ammonia. Next to blood, soybean hay produced the greatest effect; green rape was next and oat straw last. Thus it seems that nitrogenous materials, by their production of ammonia and perhaps by other reactions, reduce the lime requirement of soils. The effect has been more marked and consistent on this rather heavy soil than on the lighter soils previously studied. Carbohydrate materials have much smaller effects.

It is shown also that the limed soils have a capacity for decomposing lime-stone, even after 18 weeks' standing with an excess of lime. This would indicate that acid soils react with carbonate of lime beyond the neutral point, or that for lack of sufficiently intimate contact, all the acids have not yet been neutralized. There is perhaps no such thing as completion of the reaction. There are doubtless always soluble acids or acid salts capable of decomposing the carbonate.

TABLE 4

Lime requirement of variously treated soils (tons per 2,000,000 pounds)

	FIRST SAMPLE, 2 WEEKS	SECOND SAMPLE, 5 WEEKS	THIRD SAMPLE, 10 WEEKS	FOURTH SAMPLE, 18 WEEKS	MORE OR LESS THAN SOIL ALONE			
					First sample	Second sample	Third sample	Fourth sample
	tons	*tons*	*tons*	*tons*	*tons*	*tons*	*tons*	*tons*
Clay soil:								
Soil alone..................	3.35	2.95	3.10	3.10				
Soil limed..................	0.95	0.55	0.45	0.65				
Soybean hay...............	2.00	2.60	2.60	3.10	−1.35	−0.35	−0.50	+0.00
Limed.....................	0.35	0.95	0.60	0.80	−0.60	+0.40	+0.15	+0.15
Green rape.................	2.10	2.50	2.65	3.20	−1.25	−0.45	−0.45	+0.10
Limed.....................	0.60	0.80	0.55	0.90	−0.35	+0.25	+0.10	+0.25
Green soybeans.............	1.85	2.85	2.65	3.00	−1.50	−0.10	−0.45	−0.10
Limed.....................	0.45	0.95	0.55	0.60	−0.50	+0.40	+0.10	−0.05
Oat straw..................	2.65	2.60	2.65	2.85	−0.70	−0.35	−0.45	−0.25
Limed.....................	0.45	0.75	0.75	0.65	−0.50	+0.30	+0.30	+0.00
Blood.....................	2.00	1.80	2.00	2.95	−1.35	−1.15	−1.10	−0.15
Limed.....................	0.35	0.90	0.65	1.35	−0.60	+0.35	+0.20	+0.70
Blood and 5 tons of straw....	1.85	2.65	1.90	3.05	−0.50	−0.30	−1.20	−0.05
Limed.....................	0.35	1.05	0.80	1.50	−0.60	+0.50	+0.35	+0.85
Blood and 10 tons of straw...	1.70	2.20	2.05	3.00	−0.65	−0.75	−1.05	−0.10
Limed.....................	0.40	1.20	0.90	1.40	−0.55	+0.65	+0.45	+0.75

It is worthy of note, too, that the organic treatments seem to have increased the capacity of the soil to react with lime, when they were used alone.

There is a rather close correlation between changes in soil reaction, and the nitrogen changes as shown by table 5. This is especially noticeable on the blood treatments where there is sufficient nitrogen added to produce a measurable effect upon the reaction.

These data show a close correlation between the excess of ammonia over nitrates and the true acidity, or pH values of the soils, and would signify that the bacteriological changes which were occurring were affecting the soil reaction to an appreciable extent.

The same thing is shown in table 6 on all treatments, considering the summarized effects as before.

TABLE 5

Difference of ammonia and nitrates on unlimed soils compared with the effect of the treatment on soil reactions

SILT LOAM SOIL	FIRST SAMPLE	SECOND SAMPLE	THIRD SAMPLE	FOURTH SAMPLE
Blood treatment only				
Ammonia (p.p.m.)....·.....................	270.0	462.0	390.0	378.0
Nitrates (p.p.m.).........................	7.8	33.8	155.4	366.3
Excess (NH_3) (p.p.m.)......................	+262.2	+428.0	+234.6	+ 11.7
pH on blood..............................	6.33	7.00	6.46	5.41
pH increase over untreated soil...............	+1.62	+2.12	+1.68	+0.54

TABLE 6

Nitrogen changes and the effect on soil reaction summarized

ALL TREATMENTS NO LIME	FIRST SAMPLE	SECOND SAMPLE	THIRD SAMPLE	FOURTH SAMPLE
Ammonia (p.p.m.).........................	169.4	247.9	190.1	179.5
Nitrates (p.p.m.)..........................	26.7	88.2	109.8	242.3
Difference (p.p.m.)........................	+142.7	+159.7	+80.3	−62.8
pH values...............................	6.01	6.23	5.97	5.17
pH increase over untreated soil...............	+1.10	+1.35	+1.19	+0.29

TABLE 7

Residual carbonates on treated soils at the various samplings, expressed as tons per acre

	FIRST SAMPLE	SECOND SAMPLE	THIRD SAMPLE	FOURTH SAMPLE	MORE OR LESS THAN SOIL ALONE			
					First sample	Second sample	Third sample	Fourth sample
	tons	tons	tons	tons	tons	tons	tons	tons
Clay soil:								
Soil alone....................	2.35	2.20	1.40	0.90				
Soybean hay.................	3.15	2.40	1.05	1.05	+0.60	+0.20	−0.35	+0.15
Green rape..................	3.15	2.10	1.20	0.45	+0.60	−0.10	−0.20	−0.45
Green soybeans..............	3.25	2.00	1.45	1.00	+0.70	−0.20	+0.05	+0.10
Oat straw...................	3.20	1.85	1.40	0.95	+0.65	−0.35	+0.00	+0.05
Blood.......................	3.15	3.05	0.60	0.00	+0.60	+0.85	−0.80	−0.90
Blood and 5 tons of straw....	3.70	3.20	1.05	0.10	+1.15	+1.00	−0.35	−0.90
Blood and 10 tons of straw...	3.60	3.20	0.70	0.00	+1.05	+1.00	−0.70	−0.90

RESIDUAL CARBONATES

The data show that the organic matter protected the carbonates until there was considerable nitrification. All organic treatments caused a marked saving of carbonates at the first sampling. At the last sampling those treatments where there was much nitrogen to produce nitric acid, nearly or completely exhausted the carbonates present. Even the untreated soil reacted slowly and continually and would perhaps have used up all the limestone after sufficient time, even though there was no leaching.

These data would indicate that excessive nitrification might become a positive factor in contributing to soil acidity. However, nitrates, being soluble, will not accumulate and in the process of leaching basic material is permanently removed from the soil.

HYDROGEN-ION CONCENTRATION

The hydrogen-ion concentration was determined at each sampling on all of the treatments with the hydrogen electrode apparatus.

TABLE 8
Hydrogen-ion concentration

	FIRST SAMPLE	SECOND SAMPLE	THIRD SAMPLE	FOURTH SAMPLE	MORE OR LESS THAN SOIL ALONE			
					First sample	Second sample	Third sample	Fourth sample
	pH	pH	pH	pH	pH	pH	pH	pH
Clay soil.....................	4.91	4.88	4.78	4.88				
Soil alone:								
Soil and 5 tons of lime.......	7.62	7.72	7.65	7.60				
Soybeans and straw..........	6.03	5.89	5.50	5.02	+1.12	+1.01	+0.72	+0.14
Soybeans and lime..........	7.74	7.64	7.41	7.51	+0.12	−0.08	−0.24	−0.03
Green rape..................	6.03	6.05	6.17	4.78	+1.12	+1.17	+1.39	−0.10
Green rape and lime..........	7.66	7.65	7.42	7.53	+0.04	−0.07	−0.23	−0.07
Green soybeans..............	5.81	5.58	5.34	5.17	+0.90	+0.70	+0.56	+0.29
Green soybeans and lime.....	7.74	7.64	7.60	7.74	+0.12	−0.08	−0.05	+0.14
Oat straw..................	5.21	5.07	5.41	5.00	+0.30	+0.19	+0.63	+0.12
Oat straw and lime..........	7.48	7.60	7.66	7.71	−0.14	−0.12	+0.01	+0.11
Blood......................	6.48	7.17	6.55	5.43	+1.57	+2.29	+1.77	+0.55
Blood and lime..............	7.71	7.91	7.22	7.60	+0.09	+0.19	−0.43	+0.00
Blood and 5 tons of straw.....	6.28	7.10	6.58	5.38	+1.37	+2.22	+1.80	+0.50
Blood, 5 tons of straw and lime	7.74	7.76	7.34	7.54	+0.12	+0.04	−0.31	−0.06
Blood and 10 tons of straw....	6.24	6.74	6.24	5.44	+1.33	+1.86	+1.46	+0.56
Blood, 10 tons of straw and lime	7.65	7.76	7.36	7.61	+0.03	+0.04	−0.29	+0.01

The lime requirement according to the Tacke method was a little more than 3 tons. To take care of acids which might be produced in the decomposition of organic material, an excess of 2 tons was used. The data (table 8) show that this was sufficient to give a slightly alkaline soil either with or without organic treatment (the smaller the pH value the more acid the soil). Every organic treatment without lime diminished the true acidity of the soil, the highly nitrogenous materials most, as was true also of the lime requirement. The oat straw had the least effect. In the presence of lime, however, the organic treatments had a rather slight effect in reducing the hydrogen-ion concentration at first, and by the third sampling the effect was the reverse in nearly every case, though again the increase in hydrogen-ion concentrations was not large. By the fourth sampling the effects were quite erratic. In nearly every case where lime was not used, however, the organic treatments reduced the acidity somewhat.

GENERAL DISCUSSION

The materials used in this study were such as are common crop residues or fertilizers. The nitrogen content was: oat straw, 1.05; green soybeans, 2.41; green rape, 3.43; soybean hay, 6.63; and dried blood, 13.93 per cent. The 5-ton application of limestone proved to be scarcely enough to take care of the natural soil acidity plus that produced in nitrification as shown by the data.

The lime requirement shown by the Tacke method on this soil was about 3 tons. Shaking and aeration was continued for only 5 hours, however, in this and the remaining work, partly for convenience and partly because of the fact that a limited amount of work had shown that the lime requirement indicated by a 5-hour run was sufficient. When that quantity of lime was added to the soil and allowed to stand for a short time with optimum moisture conditions, a practically neutral reaction was shown by hydrogen-ion determinations.

The results of the effect of carbohydrate materials upon nitrification have a practical bearing which is worthy of consideration. Experience has shown that the plowing under of green manures such as rye, the heavy use of straw, and other refuse, often cause disappointing yields from the crop immediately following. This may result not alone because the crop has exhausted the water supply previous to plowing under, but oftentimes no doubt, because such materials have furnished the soil organisms with easily available sources of energy, and nitrification does not proceed rapidly enough to supply the crop with nitrates. Thus the immediate crop suffers nitrogen starvation, though perhaps later crops might be much benefited.

SUMMARY

1. Oat straw again reduced nitrification and ammonification below that of the untreated soil.

2. A mixture of straw and blood reduced the total nitrogen found in the form of ammonia and nitrates below that of the blood treatment alone. Ten tons of straw with the blood caused a somewhat greater reduction than the 5-ton application.

3. All the treatments reduced the lime requirement indicated by the Tacke method, until nitrification had taken place.

4. Lime-requirement determinations of the limed soils showed that the treated soils were always capable of reaction with more lime, though an excess of 2 tons of limestone had been applied. This shows that the soils

contain acids which are very slowly reactive, and perhaps they will react with limestone beyond their neutral point.

5. The residual carbonates, where blood was applied, were completely exhausted at the last sampling.

6. The hydrogen-ion determinations show that in practically every case the organic treatments reduced the true acidity. In some cases, on the contrary, both lime and organic treatments did not give as alkaline a soil as did the lime alone.

7. Changes in soil reaction especially on the blood-treated soils, follow very closely the deficit or excess of ammonia over nitric nitrogen, indicating that these processes may become factors influencing the production of acid soils.

BUFFERING IN SOILS

Practically all soils possess perhaps some degree of buffering, that is, they are able to react with either base or alkali to a certain extent, without very

TABLE 9

Table showing treatments and the hydrogen-ion concentration increments corresponding

SOIL	ALONE	1000 POUNDS Ca(OH)$_2$	2000 POUNDS Ca(OH)$_2$	4000 POUNDS Ca(OH)$_2$	8000 POUNDS Ca(OH)$_2$	16,000 POUNDS Ca(OH)$_2$	32,000 POUNDS Ca(OH)$_2$
	pH	pH	pH	pH	pH	pH	pH
A. Muck..............	4.5	0.2	0.3	0.2	0.6	0.3	0.7
B. Fine sand.........	5.1	0.1	0.3	0.7	1.0	1.0	0.1
C. Red clay..........	5.0	0.2	0.2	0.7	0.8	0.7	0.9
D. Coarse sand.......	5.8	0.6	0.9	0.5	0.4	0.3	0.8
E. Mucky loam.......	4.6	0.1	0.1	0.3	0.9	1.2	0.5
F. Neutral soil........	7.0	0.3	0.3	0.3	0.3	0.2	
G. Alkaline soil.......	7.9	0.3	0.0	0.0	0.2	0.1	

much change in hydrogen-ion concentration. The degree of buffering and the rate of change of reaction with increasing amounts of base or acid will depend very much upon soil type, as the following data will show.

Five grams each of the soils listed in table 9 were treated with 0.02 N Ca(OH)$_2$ equivalent to the various amounts of lime per acre of 2,000,000 pounds of soil, evaporated to dryness on the steam bath, taken up with 20 cc. of water, allowed to stand over night, and the hydrogen-ion concentration determined with a hydrogen-electrode apparatus. The acid-treated soils were managed in the same way, 0.008 N H$_2$SO$_4$ being used.

The tabulated data show that the rate of change of reaction with increasing increments of lime is very different for the different soils. The muck soil shows the highest buffering and the sand the least, as would be expected. The neutral and alkaline soils do not change very greatly, showing that they have little capacity for buffering against a base.

The data in table 10 show the effect of acid treatments.

As was true of buffering against bases, the organic soils show a greater capacity for buffering against acids. The sandy soil shows less buffering, and the neutral and alkaline soils have great apparent buffering power, perhaps due to the presence of excess bases.

In general, mucky or organic soils should show the highest degree of buffering, clays less, and sands the least. The protein materials of the organic soils, and the acid silicates of clayey soils are doubtless responsible for most of the buffer action of such types. Sands, containing perhaps little of either, are not usually highly buffered.

The highly buffered soils should show not only less change with the first treatments of base or acid but should continue to resist change of reaction longer when larger treatments are given. The initial reaction, of course, will be a factor to consider at this point. But it is worthy of note that the soils *A* and *C*, which are most acid to start with, show the greatest capacity for buffering against acid.

TABLE 10

Treatments and corresponding H-ion concentrations by increments corresponding to treatments

SOIL	ALONE	1000 POUNDS H_2SO_4	2000 POUNDS H_2SO_4	4000 POUNDS H_2SO_4	8000 POUNDS H_2SO	16,000 POUNDS H_2SO_4
	pH	pH	pH	pH	pH	pH
A. Muck................	4.5	0.4	0.2	0.2	0.2	0.4
B. Fine sand............	5.1	0.3	0.3	0.5	0.5	0.4
C. Red clay.............	5.0	0.4	0.2	0.4	0.5	0.4
D. Coarse sand..........	5.7	1.0	0.4	0.8	0.4	0.3
E. Mucky loam..........	4.5	0.3	0.3	0.3	0.3	0.3
E. Neutral..............	7.0	0.1	0.1	0.3	0.3	2.0
G. Alkaline.............	7.9	0.1	0.0	0.2	0.0	0.0

It might be supposed that since soils tend naturally to become acid the capacity for buffering against acids would be more or less exhausted. It is demonstrated that this is true to a limited extent only. While the first application of acid causes a comparatively large change in reaction, it is observed that there is a marked buffering which continues to be manifested with the highest treatments. The acid soils likewise, however, have a greater capacity for base buffering.

These facts are best brought out by means of graphs (fig. 1), which show the rate of change by the degree of curvature. Soil *A* has a curve much less steep than the other soils, soil D having much the most abrupt slope. The acid curves for *B* and D reach a final point nearly together, though starting quite widely separated. Soil *A*, which is by far the most acid, never rises to as high an acidity as soil D, which is by far the least acid. Soil *A* is a muck, while soil *D* is a sand and this difference in buffering capacity could be predicted, though such an extreme effect may seem extraordinary.

The above data have considerable significance in various ways. They demonstrate what practical experience has already indicated, that soils may be quite acid when the total lime requirement is measured, and yet have a comparatively low active acidity. Ordinary soil-acidity methods measure the capacity of the soil for decomposing lime rather than its true acidity or hydrogen-ion concentration. Soils high in organic matter may be able to take up large amounts of limestone, when a great part of this acidity has been overshadowed by amphoteric substances.

Highly buffered soils also may permit vigorous bacterial activity, because the buffering effect keeps down the hydrogen-ion concentration to a point

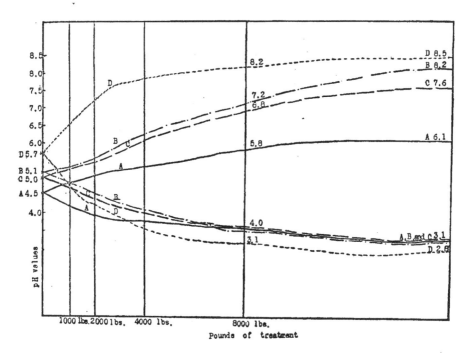

FIG. 1. TITRATION CURVES FOR SOILS A, B, C, AND D, WITH DIFFERENT AMOUNTS OF
Ca(OH)₂ AND H₂SO₄ ADDED
Ca(OH)₂ Curve—upper graph
H₂SO₄ Curve—lower graph

which is not destructive of the soil organisms. A soil on the other hand which is not buffered, has a higher hydrogen-ion concentration though a smaller total lime requirement, and organisms are not very active because of the deleterious effects of the unbuffered acids.

The importance of hydrogen-ion concentration biologically may be shown by the following data taken from Fred's work (2) with legume bacteria. Only the acid limits are given, but perhaps the alkaline limit would be nearly as far above neutrality, which would mean a wide variation for some organisms and only a narrow one for others.

Many soil organisms are even more sensitive to reaction than some of the common legume organisms, and thus the true acidity of soils is doubtless the determining factor for the biological changes which are to occur. No data can be given here for the reaction permitting mold growth, but it is known that they endure high degrees of acidity and probably no soil under ordinary treatment is ever too acid for their activity.

THE NATURE OF SOIL ACIDITY

Hydrogen-ion studies

In the following brief study tumblers of soil treated in various ways were used to determine the effect of the treatment upon the hydrogen-ion concentration. In each case 100 gm. of dry soil were employed, and the moisture content kept at the optimum (50 per cent of saturation). One series was treated with ammonium sulfate at the rate of 1 ton per acre, and lime in increasing increments, 1, 3, 5, 7, 9, 12, and 20 tons per acre of 2,000,000 pounds of soil. The results are given in table 11 in pH values.

The lime requirement of the untreated soil determined by a 5-hour Tacke run was 3.2 tons per acre. It will be observed that the ammonium sulfate alone increased the acidity, as would be expected of a physiologically acid salt which has been nitrified. The increased acidity is not overcome by the 1-ton treatment of calcium carbonate, but is more than overcome by the 3-ton treatment. A neutral reaction is not secured however, until 5 tons are applied when it runs beyond neutrality. After 9 tons are applied there is only a small increase in alkalinity, and with 20 tons the pH is not quite 8.

In table 12 similar results are presented from the tests with an organic nitrogenous material, albumin, applied at the rate of $1\frac{1}{2}$ tons, or approximately the equivalent in nitrogen content of the 1 ton of ammonium sulfate.

The results are very similar to those obtained with ammonium sulfate. Accidentally or otherwise, the albumin caused a slightly greater acidity when lime was not applied to the soil but in most cases it was less. In other words, the same amount of lime permitted less acidity or more alkalinity when albumin was used than when an equivalent amount of ammonium sulfate was used. This may be due to the fact that not only was nitric acid produced from ammonium sulfate but sulfuric acid also remained. When albumin was nitrified if any other acid was produced it was in a smaller quantity or more slightly ionized than the sulfuric acid from the ammonium sulfate.

TABLE 11

The hydrogen-ion concentration values for the various treatments, incubated 6 weeks

	SOIL ONLY	SOIL (NH₄)₂SO₄	SOIL (NH₄)₂SO₄	SOIL (NH₄)₂SO₄	SOIL (NH₄)₂SO₄	SOIL (NH₄)₂SO₄	SOIL (NH₄)₂SO₄	SOIL (NH₄)₂SO₄	SOIL (NH₄)₂SO₄
Lime	0	0	1 ton	3 tons	5 tons	7 tons	9 tons	12 tons	20 tons
pH	5.08	4.88	4.98	5.67	7.27	7.86	7.91	7.91	7.96

TABLE 12

Hydrogen-ion concentration values with albumin treatments

	SOIL ONLY	SOIL ALBU-MIN	SOIL ALBU-MIN	SOIL ALBU-MIN	SOIL ALBU-MIN	SOIL ALBU-MIN	SOIL ALBU-MIN	SOIL ALBU-MIN	SOIL ALBU-MIN
Lime	0	0	1 ton	3 tons	5 tons	7 tons	9 tons	12 tons	20 tons
pH	5.08	4.76	5.27	6.07	7.84	7.96	7.91	7.91	8.02

TABLE 13

Hydrogen-ion concentration values with various lime applications on soil alone

	SOIL	SOIL	SOIL	SOIL	SOIL	SOIL	SOIL	SOIL
Lime	0	1 ton	3 tons	5 tons	7 tons	9 tons	12 tons	20 tons
pH	4.70	4.91	6.55	7.69	7.68	7.90	8.05	8.26

TABLE 14

Hydrogen-ion concentration values of soils treated with acids and varying amounts of lime

	SOIL ONLY	ACIDS AND SOIL	ACIDS AND SOIL	ACIDS AND SOIL	ACIDS AND SOIL	ACIDS AND SOIL	ACIDS AND SOIL	ACIDS AND SOIL	ACIDS AND SOIL
Lime	0	0	1 ton	3 tons	5 tons	7 tons	9 tons	12 tons	20 tons
pH	4.71	3.91	4.31	6.15	7.25	7.55	7.55	7.69	7.79

TABLE 15

Hydrogen-ion concentration values on soils variously treated as shown

AMMONIUM SULPHATE					ALBUMIN			
Soil	Soil	Soil	Soil	Soil	Soil	Soil	Soil	Soil
(NH₄)₂SO₄ only	H₂SO₄ 1 ton	H₂SO₄ 3 tons	H₂SO₄ 5 tons	Citric acid 10 tons	H₂SO₄ 3 tons	H₂SO₄ 5 tons	Citric acid 7 tons	Albumin alone
4.98	4.21	3.85	3.62	5.22	4.36	3.65	4.69	4.76

TABLE 16

Hydrogen-ion concentration values on soils treated with varying amounts of citric acid

SOIL ONLY	SOIL	SOIL	SOIL	SOIL	SOIL
	Citric acid 3 tons	Citric acid 5 tons	Citric acid 7 tons	Citric acid 9 tons	Citric acid 10 tons
pH	pH	pH	pH	pH	pH
4.71	5.02	5.33	5.40	5.14	5.33

In table 13 are found the results obtained where the soil alone is given the various lime treatments and the hydrogen ion determined.

Evidently there must have been some variation in the soil as this sample seems to be more acid originally. The 3-ton treatment did not produce neutrality while the 5-ton treatment produced alkalinity. Apparently about 4 tons, or a little more than the indicated Tacke requirement, is necessary to give a neutral soil. The 20-ton treatment runs above pH = 8 which is rather alkaline for a limestone treatment.

In the next series acids were added equivalent, respectively, to the nitric and sulfuric acids which would result if the ammonium sulfate were completely nitrified.

The acids increase the acidity but the 5-ton treatment of limestone gives a somewhat alkaline soil (table 14). The higher treatments do not cause as great an alkalinity as where nothing but lime is added to the soil, even when a large excess of lime is present. Another series was treated with ammonium sulfate and mineral and organic acids. Sulfuric and citric acids were used in equivalent amounts.

It is very evident that the 10 tons of citric acid in conjunction with the ammonium sulfate did not increase the true acidity of the soil (table 15). In fact, it is much less. Neither did the 7 tons used with the albumin cause any increase. But the sulfuric acid evidently caused quite a marked increase in every case, the increase being somewhat proportional to the amount applied. The 3-ton application of sulfuric acid did not have so great an effect in the presence of albumin as with ammonium sulfate, but the 5-ton treatment had nearly as great an effect.

Another series was run in which citric acid was used on the soil alone.

It is very evident again that the organic acid has not increased the acidity of the soil, and the largest application has no more effect than the smaller ones (table 16).

These results are in accord with the contention that organic acids do not accumulate in soils under conditions favorable to crop production. It is very evident that the organic acid used here has oxidized rapidly enough to remove all cause for suspicion that ordinary acid soils might owe this characteristic to citric acid produced from the decay of organic matter. The results agree also with those of Stemple, who used citric, oxalic and acetic acids. It is possible, of course, that more stable and active organic acids than citric might be produced, and that there might be conditions when such acids would contribute to the causing of an acid soil.

SOURCE OF ORGANIC AND MINERAL ACIDS

From whence arises the acidity of ordinary agricultural soils has long been a somewhat perplexing problem. It is generally believed at the present time that most of the acidity, except perhaps that in peat and muck soils, arises from some mineral source. The leaching of bases and the consequent accumulation of acid silicates and alumino-silicates is doubtless responsible for a considerable portion of acidity. The practice of using certain commercial fertilizers, such as ammonium sulfate, has caused an acid condition of some soils. Thus the accumulation of sulfuric, hydrochloric, or nitric acids even in small amounts could cause a marked increase in the harmful effects of an acid soil, because such acids are highly ionized and would therefore give a high hydrogen-ion concentration. A small amount of such acids would undoubtedly do more injury than larger amounts of either acid silicates or organic acids.

It may be easily demonstrated that soils contain acids of very variable strengths, the more active ones reacting at once and the very slowly active ones only after a much longer period of contact with limestone and water.

TABLE 17

Lime requirement at intervals of 3 hours

SOIL NUMBER		3 HOURS	6 HOURS	9 HOURS	TOTAL
1	Loam (lbs.)............................	5000	6100	6500	6500
	Increase (lbs.)...........................		1100	500	
	Per cent of total........................	77.0	17.0	6.0	100
3	Sandy loam (lbs.).......................	3700	5100	6200	6200
	Increase (lbs.)...........................		1400	1100	
	Per cent of total........................	59.7	22.6	17.7	100
4	Sand (lbs.).............................	800	1100	1100	1100
	Increase (lbs.)...........................		300	0	
	Per cent of total........................	72.7	27.3	0	100
5	Miami silt (lbs.)........................	1800	2300	2500	2500
	Increase (lbs.)...........................		500	200	
	Per cent of total........................	72.0	20.0	8.0	100

This may be due to the fact that the acids are very slowly soluble, or it may be partly because of hydrolytic actions which take place slowly. The data in table 17 showing the varying degrees of activity of soil acids, are taken from a previous work.

These results show that from 60 to 80 per cent of the acidity based upon a total 9-hour run, reacted during the first 3 hours, while there yet remained 6 to 18 per cent to react during the last 3 hours of the run, except for the sand which was not very acid. Determinations have been conducted a much longer period than this and have been found to react slowly even after several days. One muck soil with a lime requirement of 15,200 pounds at the end of 3 hours gave a 25,400-pound requirement at the end of a 23-hour period. A soil of this type, however, is quite different from the ordinary soil, and doubtless the organic acids have a part to play in its reaction.

THE LOSS OF BASES BY SOILS

It is not presumed that soils become acid so long as they contain bases equivalent to the acids. But the question may arise, do acids increase in quantity or do bases diminish in quantity and thus leave a surplus acidity, and if either or both changes take place in what manner do they occur?

The bases such as sodium, potassium and calcium must be held originally in some chemical combination, undoubtedly with a silicate or alumino-silicate, to form a salt or acid salt. This gives a salt of a strong base and a weak acid and should therefore by hydrolysis give up a free base. That such is true has been demonstrated experimentally as shown by the data from Steiger's work (1) with various natural silicates (table 18).

TABLE 18

Alkalinity of natural silicates

NAME	FORMULA	COMBINED ALKALI	EQUIVALENT OF NaO IN SOLUTION
		per cent	*per cent*
Pectolite......	$Ca_2(SiO_3)_3NaH$	9.11	0.57
Muscovite....	$Al_3(SiO_4)_3KH_2$	10.00	0.32
Natrolite......	$Al_2(SiO_4)_3Na_2H_4$	15.79	0.30
Lintonite......	$Al_6(SiO_4)_6(CaNa_2)_3 7H_2O$	5.92	0.29
Phyogopite....	$Al(SiO_4)_3Mg_3KH_2$	9.32	0.22
Laumonite....	$Al_2SiO_4Si_3O_8Ca4H_2O$	1.00	0.18
Lepidolite.....	$KHLiAl_3(SiO_4)_3K_3Li_3(AlF_2)Al(Si_3O_8)_3$	13.00	0.18
Elaeolite......	$Al_3(SiO_4)_3Na_3$	21.17	0.16
Henlandite....	$Al_6(Si_3O_8)_6(CaNa_2)_3 16H_2O$	2.00	0.13
Orthoclase....	$KAlSi_3O_8$	16.00	0.11
Analcito......	$NaAl(SiO_3)_2 2H_2O$	14.00	0.10
Oligoclase.....	$AlNaSi_3O_8Al_2CaSi_2O_8$	9.18	0.09
Albite........	$AlNaSi_3O_8$	12.10	0.07
Wernerite.....	$Ca_4Al_6Si_6O_{25}Na_4Al_3Si_9O_{24}$	11.09	0.07
Leucite.......	$KAl(SiO_3)_2$	21.39	0.06
Stibite........	$Al_2(Si_3O_8)_2(CaNa_2) . 6H_2O$	1.00	0.05
Chabazite.....	$Al_2SiO_4Si_3O_8(CaNa_2) . 6H_2O$	7.10	0.05

These results were obtained by placing 0.5-gm. samples in 500 cc. of water and maintaining at a temperature of 70°C. for a month. It is to be expected that in the soil it might go on even more readily, since the base would be leached as liberated unless perchance it reacted with some acid or protein decomposition product to form an insoluble salt. Why all bases do not leach with about equal readiness cannot be stated, but potassium seems about the least readily leached and calcium most readily leached. When several hundred pounds of limestone may be leached out in a single year it is not strange that a soil may become rather acid and unproductive in time for that reason.

There is, therefore, nothing more logical than that with increased weathering there should come increased acidity. As long as base-rich minerals are tightly cemented together or enclosed within the interstices of a resistant granite or other mineral, they are mechanically protected and saved from waste. But they are likewise saved from any useful function in the soil either as direct plant-food or as a neutralizing agent. Virgin soils are not only more likely to contain many minerals rich in unleached bases but they contain much organic matter in the process of decay and therefore in a condition to react with, and to prevent the leaching of base. With the exhaustion of the organic matter there is the accompanying loss of base and therefore a non-productive sour soil. ·

Experimental data show that practically any type of soil may become acid. But the acidity of different soil types behaves in a different way, as may be shown also experimentally. A sandy soil is likely to become acid readily because there is not sufficient organic·matter to prevent leaching of such bases as may occur naturally or may be applied artificially.

There would probably be little acidity due to organic acids, because there would likely be very little organic matter in such a soil and because conditions would probably be very favorable to the oxidation of such organic acids as might possibly develop. Mineral acids such as the acid silicates, and the stronger sulfuric and hydrochloric acids from the application of certain fertilizers, would likely cause the injurious soil reaction. On a clay soil more acid alumino-silicates would be probable. Loam soils and those of yet higher organic content might contain organic acids, or at least organic compounds capable of combining with base. Such soils remain productive in spite of such acidity as may develop because necessary bases for plant growth have been prevented from leaching and because the organic matter itself is an important source of the essential plant-food, nitrogen. In the growth of legumes, however, it is perhaps not a question of nitrogen content, but more likely a question of reaction and a supply of mineral plant-food, including not only the bases potassium and calcium, but also phosphoric acid.

GENERAL DISCUSSION

There are many factors which influence directly or indirectly the reaction of soils. It is not alone a question of the production of acids but a question of the capacity of the soil to resist changes in reaction caused by the acids produced.

Buffering may be effected by both mineral and organic compounds. Silicates of bases would be capable of neutralizing strong acids, which is in fact a buffering effect. Some of the alumino-silicates no doubt react with either acid or base and therefore function doubly, saving base and reducing the true acidity. The amino acids and many more complex products of protein degradation react in the same manner. The ionization constants for the amino acids as either acids or bases are very low but of about equal strength, making them ideal buffers. This explains why organic soils and clayey soils should show greater power to resist changes in reaction.

Grain size is of course an important factor in determining its reaction, especially with mineral soils. The smaller the grain size the more difficult it is to prevent water-logging, and therefore the more difficult to maintain conditions favorable to the oxidation of organic acids or other harmful products. Though coarse-grained soils readily become acid it is perhaps usually with a somewhat different type of acidity. Rahn (3) has already demonstrated the close relationship between grain size, moisture content, and bacterial activity. This relationship has its influence also upon reaction changes.

SUMMARY

1. Highly organic soils and clays exhibit a high degree of buffering, while coarse sands show little of this capacity.

2. Sulfuric acid, or physiologically acid salts such as ammonium sulfate, cause a change toward increased hydrogen-ion concentration in soils. Citric acid did not increase the true acidity.

3. Ammonium sulfate caused a greater increase in acidity than did its nitrogen equivalent of albumin.

4. When nitric and sulfuric acids were added to the soils in amounts equivalent to the acidity which might be produced from the complete nitrification of ammonium sulfate, a greater increase was produced in the hydrogen-ion concentration of the soil than where the ammonium sulfate was used.

5. A large excess of pure lime carbonate (20 tons) brought the pH value to only a little more than 8.0, which seems to be about the limit of alkalinity produced by limestone.

REFERENCES

(1) CLARK, F. W. 1900 Contributions to chemistry and mineralogy. U. S. Geol. Survey Bul. 167, p. 156.

(2) FRED, E. B., AND DAVENPORT, AUDREY 1918 Influence of reaction on nitrogen-assimilating bacteria. In Jour. Agr. Res., v. 14, no. 8, p. 317.

(3) RAHN, OTTO 1912 The bacterial activity in soils as a function of grain size and moisture content. Mich. Agr. Exp. Sta. Tech. Bul. 16.

(4) STEPHENSON, R. E. 1919 The effect of organic matter on soil reaction. In Soil Sci., v. 6, p. 413–439.

(5) STEPHENSON, R. E. 1921 Soil acidity and bacterial activity. In Soil Sci., v. 11, p. 133–144.